Understanding Headspace

#2 in the Gunsmithing Student Handbook Series

By Fred Zeglin

Understanding Headspace

Copyright© 2014, Fred Zeglin

All rights reserved. No part of this book may be reproduced or transmitted in any form or by any means, electronic or mechanical, including photocopying, recording or by any information storage and retrieval system, without written permission of the author.

ISBN# 978-0-9831598-4-1

Library of Congress
Control Number 2016905501

Published by
4D Reamer Rentals LTD.
432 E. Idaho St., Suite C420
Kalispell, MT 59901
www.4drentals.com

Foreword

I have known Fred Zeglin for several years. He is a highly competent, intelligent, gunsmith, instructor, author and friend.

As a gunsmithing instructor myself, it is wonderful to see something so misunderstood as "headspace", be covered in the concise easily understood format as Fred has put together here. It is nice to see Fred's humor come out in some of his writing as well.......since it appears too many writers/instructors take themselves too seriously these days.

Thanks for putting together an easy to read and understand text.

Ken Brooks
Owner/Gunsmith
PISCO
140 East Third Street
Coquille, OR 97423

www.piscogunsmithing.com

Acknowledgments

Dave Kiff of Pacific Tool & Gauge
Robert Dunlap, Master Gunsmith
Speedy Gonzalez, Member U.S. Benchrest Hall of Fame

Special Thanks to Sporting Arms and Ammunition Manufacturers' Institute, Inc. (SAAMI) for the use of their prints in this booklet.
http://www.saami.org

Preface

Recently it came to this author's attention that at least one gunsmithing school has attempted teaching barrel installation as part of their CNC classes. They are not teaching about headspace or headspace gauges. Apparently somebody has the idea that technology now supersedes the need for understanding, or for that matter the proper use of tools.

If all gunsmiths were given a CNC machine at graduation this system might work, however, they would still lack the knowledge necessary to diagnose problems that relate to various headspace conditions.

In preparation of this booklet the author naturally read everything possible on the subject of headspace in an effort to distill all the salient information into one location for the gunsmithing student. Many self proclaimed experts try to create fear and confusion around the subject of headspace. This subject, like most measuring processes, is simple and easy to understand if you are willing to educate yourself.

A basic understanding of headspace and how it affects the proper operation of a firearm are necessary if you wish to call yourself a gunsmith. Otherwise all you will be able to do is change parts with the hope that somehow the problem will go away. The later is normally called an armorer, not a gunsmith.

This booklet contains all the information a gunsmith needs in order to understand and diagnose headspace.

TABLE OF CONTENTS

PREFACE 5

ACKNOWLEDGMENTS 6

WHAT IS HEADSPACE? 11

HEADSPACE METHODS 12
   Rimless, Rebated or Semi-rimless Cartridges 12
   Rimmed Cartridges 15
   Belted Cartridges 16
   Rimless Pistol Cartridges 18

HEADSPACE AND PRESSURE 20

ACKLEY IMPROVED CARTRIDGES 26
   Rimless or Rebated Improved Chambers… 26
   Chambering an Improved Rimmed Case… 28
   Chambering an Improved Belted Chamber… 29

TO GAUGE OR NOT TO GAUGE? 32

STANDARDS AND TOLERANCES… 37

SHIMMING GAUGES 40

HEADSPACE INTERCHANGE CHART 41

ABOUT THE AUTHOR 46

What is Headspace?

In a nut shell, Headspace is the distance from the breech face to the datum line of the cartridge when chambered.

In a broad sense, headspace is a complex relationship between the cartridge, chamber, and firearm mechanism. The starting point for headspace is the cartridge case. The complex functions of the cartridge case include holding the cartridge components together, aligning the bullet in the bore, expanding to seal the breach from gasses, contracting in time for easy extraction, and removing heat from the chamber. Proper space for the cartridge allows it to function well. Improper space can cause dangerous pressure conditions at worst, or interfere with normal function of the firearm at a minimum.

There are a number of popular misconceptions concerning headspace, most notably:

- the tighter the headspace, the better.
- that loose headspace is dangerous.
- and that one particular dimension is best.

None of these common beliefs is totally true. Insufficient or unnecessarily tight headspace tends to cause malfunctions such as failure to lock. Tight headspace often makes the action stiff and hard to function. It can make extraction difficult and can cause dangerous stresses on the mechanism that shorten its life expectancy or lead to failure of the firearm.

Excessive headspace may lead to gas leakage around the case, head separation or on rare occasions sudden release of high-pressure gas. The release of gas would not be from excessive pressure in such a situation but from the failure of the unsupported case. Excessive headspace promotes wear on the firearm. Most shooters fear excessive headspace, but it is actually insufficient headspace that causes more problems. A good firearm design can actually tolerate a great deal of excess headspace. [1]

Correct headspace for any specific firearm will produce reliable function, long life (of both the brass and firearm), and good accuracy. Some firearms require a little headspace to function correctly, for example revolvers must have .002" headspace so that the cylinder can rotate smoothly.

Bolt actions will deliver best accuracy when they're headspaced with zero (.000") headspace. It's up to you the gunsmith to know what is correct for the gun at hand.

Headspace Methods

There are four basic methods of measuring headspace in firearms. We will address each method in the pages that follow.

Rimless, Rebated or Semi-rimless Cartridges (Bottleneck cases)

    1. Rimless, semi-rimless and rebated cases utilize the "Datum Line". Many sources treat the semi-rimless as a separate type of headspace. This is incorrect, as semi-rimless cases utilizes the datum line on the shoulder just like their bothers the rimless and rebated cases.

    This datum line method of measurement refers to a specified point on the shoulder of the cartridge that is a predetermined diameter. Headspace is measured from the bolt face to this predetermined point called the datum line. By way of example the 270 Winchester's datum line is the point along the shoulder of the chamber or gauge that measures .375" in diameter.

---

[1] Armalite, Inc., Technical Note 69, 2008

Cartridges that are registered with SAAMI will have dimensions specified on prints for that case and approved by SAAMI. Wildcats can and should have a datum line too.

A Go gauge is set to Minimum dimensions for the cartridge. Most No-go gauges are .004" longer than the Go, although they may vary slightly. A Field gauge is normally set to SAAMI maximum length for that cartridge. Technically any gun that does not accept the Field gauge is still safe to fire. That condition is for simple function, it does not address accuracy and longevity of the firearm.

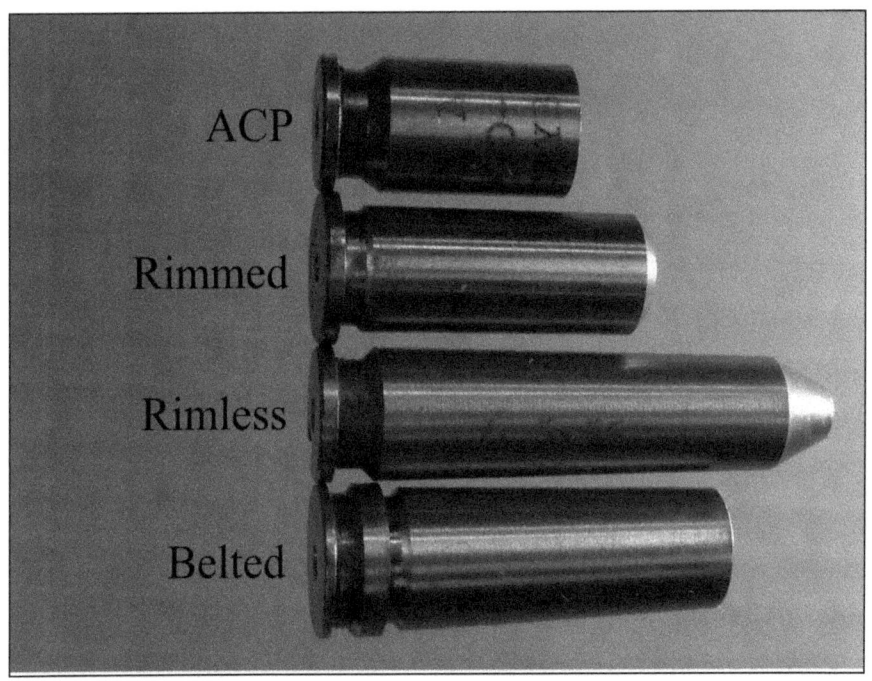

Pictured here are the common types of headspace gauges. Headspace gauges are really chamber length gauges that provide a reliable and easy way to check the headspace of any given firearm.

Note that in the drawing on the next page, SAAMI calls out a Maximum and Minimum headspace limit. In this case .010". They do not specify Go, No-go, and Field dimensions. These are dimensions set forth by the tool makers in the industry based on SAAMI standards by further breaking down the dimensions to be more useful to the gunsmith and shooter alike.

⊗ *This symbol is used to denote headspace dimensions on SAAMI prints.*

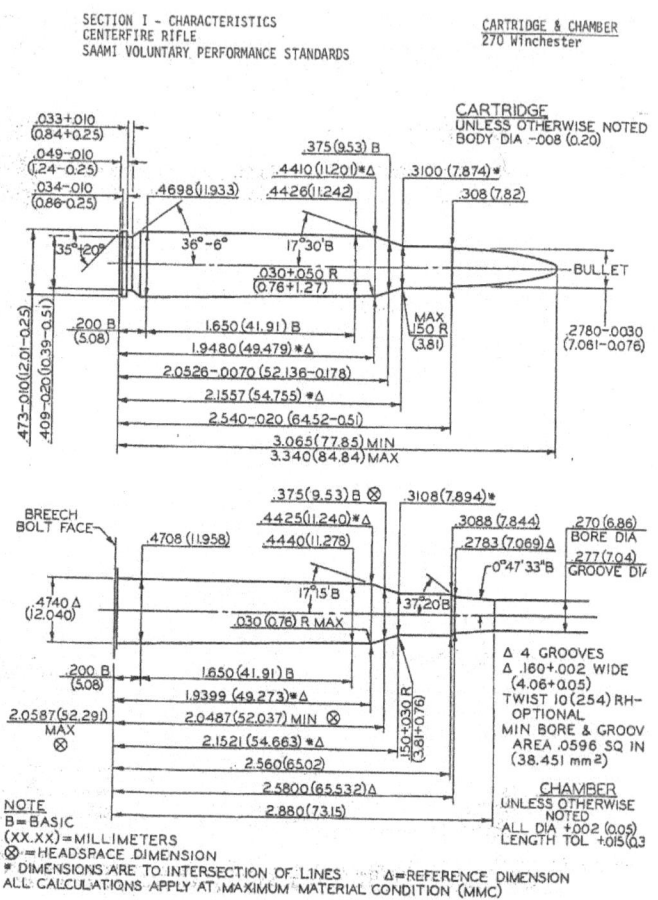

Rimless Example

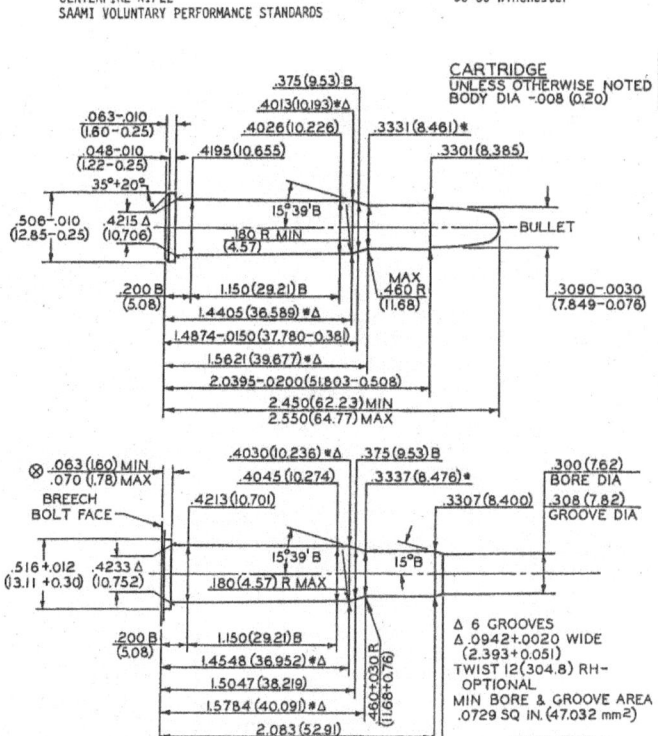

Rimmed Example

Rimmed Cartridges

2. Rimmed cases are headspaced by the thickness of the rim. The distance from the bolt face to the front edge of the cartridge rim is the headspace.

The forward edge of the case rim should fully contact the back of the barrel when the bolt is in the closed position. Rim thicknesses vary, with the majority ranging from .060" to .070" with today's cartridges. Gauges for these calibers are normally stubby and do not match the chamber itself beyond the rim.

Rimmed cases according to SAAMI tend to have a tighter maximum headspace of only .007". Most are low pressure cartridges and the primary reason for tight headspace is for best accuracy and longevity of the firearm.

While we always try to keep headspace near zero (in terms of excess headspace) there are some guns which must have a little headspace to function. For example, lever actions usually function better with a tiny amount of headspace (about .001"). Set headspace too tight and such actions must try to crush the solid brass of the rim in order to fully close.

Belted Cartridges

3. Belted cases measure headspace from the face of a fully locked bolt to the front edge of the belt. For standard magnums based on the H&H case this measurement is .220"

    At the time of this writing the only exceptions to this system are the 240 Weatherby which is a belted 30-06 case, according to SAAMI it headspaces at .219". The other exception is the .378 or .460 Weatherby case head, this family of cases headspace at .252". Finally, the 450 Marlin has a specific headspace gauge measuring .252" but with a smaller diameter than the Weatherby .460.

    Belted gauges, like rimmed, are stubby and do not match the specific cartridge. Belted gauges are only checking the depth of the belt cut in the chamber, not the length to the shoulder datum.

SECTION I - CHARACTERISTICS
CENTERFIRE RIFLE
SAAMI VOLUNTARY PERFORMANCE STANDARDS

CARTRIDGE & CHAMBER
300 H&H Magnum

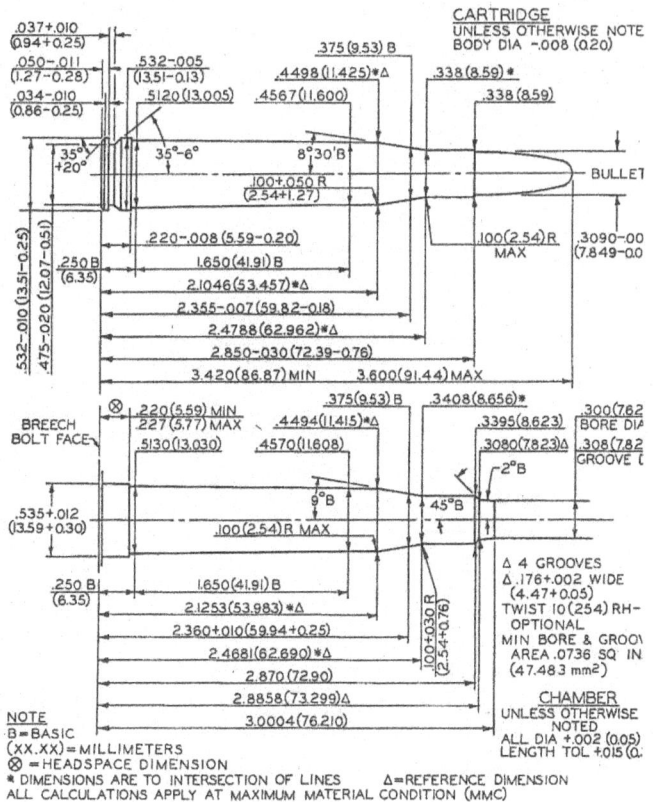

Belted Example

Magnum cartridges normally have a maximum headspace of .007" this is in part because they operate at higher average pressures than most non-magnum cartridges. It is much easier to damage a chamber with a belted gauge because of it's small contact area with the chamber and the

17

relatively large amount of leverage available to the person closing the bolt. So be gentle.

Benchrest gunsmith and Benchrest Hall of Famer, Speedy Gonzalez mentioned that when he headspaces belted magnums he uses custom gauges that are made to measure to the datum on the shoulder. This means he needs a gauge for each cartridge instead of simply using the belted gauges. I.E. you would need separate gauges for 300 Winchester Magnum and 338 Winchester magnum.

Speedy reports that it's easier to gauge the chamber with full length gauges and that accuracy often improves because the gauges are less susceptible to dirt etc. Speedy has experience teaching gunsmithing as well and he found that students may not realize how important it is that the gauges be clean. Belted gauges can easily collect dirt, oil, or chips in front of the belt. The belt cut in the chamber must also be clean of dirt, chips and oil when checking headspace.

To be clear, such full length gauges are not the standard in the industry and would be custom order. Many shooters use the method of headspacing their ammunition to the shoulder. They accomplish this by being careful not the move the shoulder during the resizing process. Most resize dies are capable of moving the shoulder by .005" or more. Learn how to use reloading dies correctly.

Rimless Pistol Cartridges

4. Rimless pistol cartridges are usually straight walled, with no shoulder or rim to headspace on. Headspace on such cartridges is measured from the fully locked breach face to the mouth of the case. The best known example of this is the 45 ACP.

Now that we have discussed the various methods of measuring headspace, it's important to realize that whatever the method, headspace must be set within specific limits to allow proper function of the firearm.

In most guns headspace should be near "0" (zero). What we mean by that is that there should be no gap between the bolt face and the back to the gauge in the chamber when the gun is locked up.

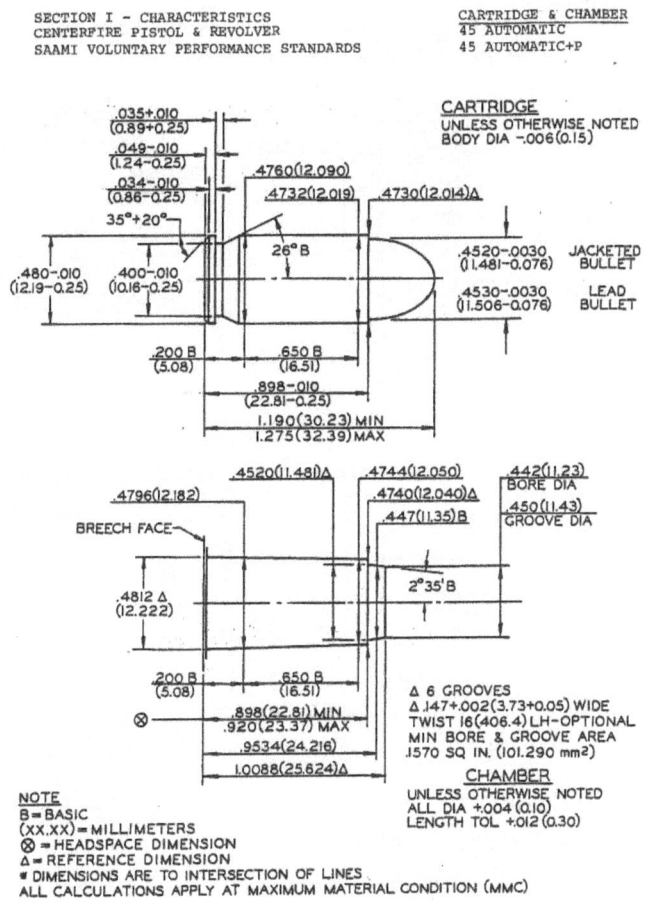

Rimless (ACP) Pistol Example

To clarify: The headspace gauge in whatever form is simply a length gauge for the specified length for that chamber. True headspace measurement is the distance + or − the length of the gauge, from the breech face to the datum line. Most commonly we concern ourselves with excessive headspace, where the chamber is longer (or +) the length of the gauge. Think of the headspace gauge as a "chamber length gauge" and this whole discussion is simplified.

The firearms industry is changing rapidly, today we can have short run custom brass or ammo made for any wildcat with proper headstamps, something that was totally cost prohibitive in the past. So, it's more important than ever to "standardize" improved and wildcat designs. The liability associated with building rifles with non-standard headspace is growing with our society's penchant for lawsuits. Always mark your barrel jobs with the correct name to minimize liability.

P.O. Ackley's cure for this problem was to supply dies with the rifle when he delivered it, a solution that worked for him. In this way he eliminated the problem of the client getting the wrong dies and not understanding why they do not work correctly.

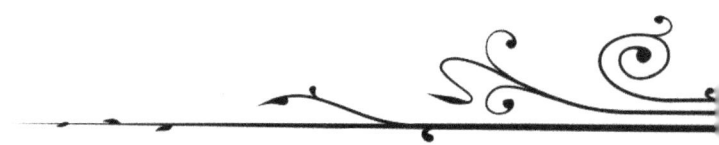

Headspace and Pressure

Shooters and some gunsmiths have a hard time discerning between headspace problems and pressure problems.

It's easy to check headspace, simply grab the correct gauges and check the gun. There are potential ways to get a false reading from a gauge. Extractors, ejectors and firing pins can all interact with the gauge to give false readings. Best results are achieved with all of these removed during headspace checks. One rare problem can occur if the gauge rim is too large for the breach face (make sure the gauge is seated fully against the breach face). Also if there is a rim recess cut in the barrel, make sure the recess is clean the gauge can seat to the bottom of the rim recess.

Once you know that headspace is correct then move on to locating the real cause of reported problems. All too often reloaders cause the problems that they call headspace with poor reloading practices or a lack of understanding of how ammunition interacts with the chamber.

Excessive pressure can be caused by many different things. Among the top most common issues are:

- *Hot loads, too much powder*
- *Choice of powder, inappropriate to the cartridge*
- *Wrong primer combined with load*
- *Short throat in chamber*
- *Tight chamber dimensions, especially neck (non-standard)*
- *Bullets with unusually long contact surfaces*
- *Bullet jammed into the lands with no jump*
- *Brass too long for the chamber (case mouth pinched by end of chamber)*
- *Mixing components without "working up" the load properly*
- *Wrong Cartridge for chamber (don't laugh, this happens often)*

Excessive headspace does not cause pressure and will not "blow up" a gun. Excessive headspace can lead to case head separations which are often misunderstood as pressure related. Case head separations are caused by excessive headspace, which allows the brass to stretch so much that it is pulled apart... Fix the headspace and the head separations will stop.

Gunsmithing Axiom:
There is one cause of case-head separations; excessive headspace.

Using Primers to diagnose headspace...

It's not unusual to have a reloader claim that his new rifle has headspace. Well of course it does, all rifles have headspace, what counts, is whether it's correct or not.

Below are three spent primers. All came from rifles with correct headspace set to zero (.000"). The primer on the *left* is what most gunsmiths and reloaders would call normal. Actually, I would consider this primer to be from a low pressure loading. It has not been expanded at all by pressure inside the case.

Center we have a primer from a factory loading. Note that the back of the primer is flattened to some degree as compared to the low pressure load on the left. There is also slight evidence of what many loaders would call cratering at the firing pin mark. This is caused in this case by the firing pin moving back slightly while the primer is under full pressure. (A heavier firing pin spring will often stop this type or cratering.)

Cratering is often caused by an oversized firing pin hole, or by a light firing pin spring that allows the pressure to force the pin back, thus flowing the primer into the firing pin hole. Excessive pressure can also cause cratering, if pressure is the culprit then the clients loads are way too hot and on the verge of a catastrophic failure. Normally with excessive pressure you will see other signs of pressure on the fired cases. If the clients loads are the culprit, when you test fire with factory loads in the same rifle the cratering likely does not happen.

The Primer on the *right* is what many reloaders and gunsmiths would call flattened. This is totally incorrect. This primer was fired with a condition of excessive headspace. The rifle has been headspaced correctly but the ammunition was made too short by the reloader who set his dies hard against the shell holder without checking the fit of the brass to the chamber.

When the round is fired in this condition the primer backs out until it hits the bolt face. Pressure is still on the rise so the primer begins to balloon as pressure rises. Then the brass stretches under pressure and the head moves back to the bolt face, in the process forcing itself down over the expanded primer. This reseating of the primer causes the flange or lip to be created as the primer conforms to the primer pocket. This is not a hot load situation; this type of primer condition is created with standard pressure loads for the cartridge.

If you see a primer such as the one on the right, you can double check the question of headspace vs. pressure very easily. First check headspace in the firearm with the correct gauge. If it is correct, check the primer pocket in the fired case if it has not expanded then the pressures were normal and you have a reseated primer.

When pressure is the culprit and you see a flattened primer there are normally many other signs of pressure. First, bolt lift is hard. Second, the case head will be expanded, thus creating a loose primer pocket. Many times you will note that the stampings in the back of the case have been closed up (flattened) substantially. The brass may have flowed into the ejector slot or hole in the bolt face. Sometimes the case head has expanded so much that a burr in lifted on the circumference of the case rim where the extractor or bolt rotated and moved the expanded brass.

Primers are a poor source of information relating to headspace. Every manufacturer uses different specifications and materials in their primer, so it is impossible to get definitive information from spent primers. However, they can provide the gunsmith useful information about the way the ammunition was loaded. Add that information to the knowledge of correct headspace and you can diagnose many problems.

Speedy Gonzalez, Custom Benchrest gunsmith, Member of the Bechrest Hall of Fame and gunsmithing instructor had some comments on primer pockets as a possible source of problems.

"It's common among reloaders and in particular benchrest shooters to uniform primer pockets in an effort to improve overall accuracy. Speedy says he frequently sees primer pockets that have been cut too deep during the uniforming process. Many of the tools used to uniform primer pockets are adjustable so it is possible to cut too deep.

Accuracy can suffer is the primer is too deep or not fully seated. The amount of energy transmitted from the firing pin to the primer is reduced if the primer is not fully seated or is too far away from the bolt face. Reduced firing pin strike caused inconsistent ignition and thus erratic pressures. The result is accuracy problems that are hard to track down. This is an example of how inspecting the client brass can help you identify a problem.

The primer being seated too deep will mimic excessive headspace in the form of light primer strikes and poor accuracy. If headspace measures correctly in the rifle but the extreme spread of velocities over the chronograph are wide, start looking for problems with ignition."

PROPERLY HEADSPACING ACKLEY IMPROVED CARTRIDGES

Headspace for Ackley Improved cases should be a no-brainer. Ackley set up probably the simplest headspace system for a line of wildcats that any gunsmith ever devised. The author is aware of numerous situations were Ackley cartridges have been incorrectly headspaced. Why does this matter? Because it can cause a catastrophic failure which will damage the gun and could permanently blind the shooter, or worse.

Rimless or Rebated Improved Chambers...

Ackley Improved cartridges in this category seem to receive the most abuse at the hands of hobbyists and local gunsmiths who do not understand the proper headspace of Ackley Improved designs. P.O. Ackley did establish specific headspace dimensions for all his Improved case designs. The process is extremely simple and for this reason alone folks seem to think they need to make it more complex. Keep it simple.

The most important innovation that Ackley brought to the "improved" concept was with regard to bottleneck rimless cases. He chose the simplest of mechanical solutions to insuring that his improved cases would safely fire factory loads. He shortened the chamber by .004" (4/1000 of an inch).

By making his headspace gauge .004" shorter; a factory case is then a crush fit between the bolt face and the junction of the neck and shoulder, proper headspace is insured. This is why Ackley prescribes setting the barrel back when rechambering for such cases.

P.O. Ackley referring to his Improved case designs, "When checking the headspace, a standard "Go" gauge with .004" ground off the head is the proper one to use. In other words, the headspace has to be minimum-minus .004" in order to prevent case head separations." So the standard "Go" gauge for the factory caliber becomes the "No-Go" gauge for the Ackley Improved chamber.

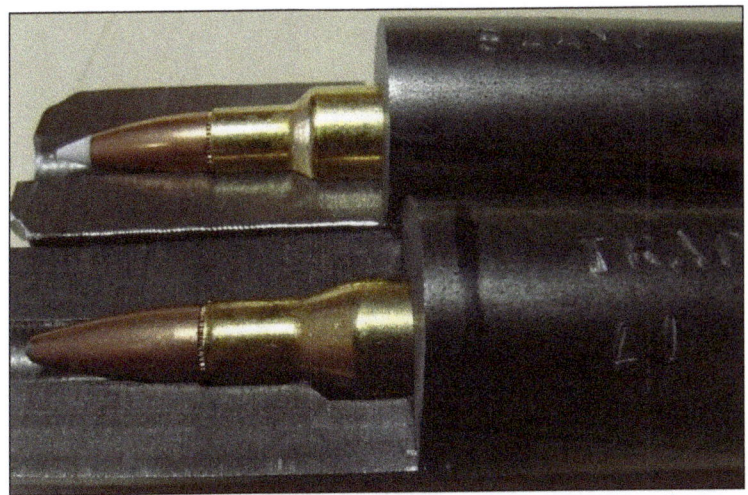

Top is a fully formed Ackley cartridge in the chamber, bottom is the same chamber with a standard factory parent case in the chamber. Note how the neck/shoulder junction interacts with the chamber to headspace the factory cartridge.

Some confusion seems to arise out of this headspace system. Folks get confuse between the set up for a rimless bottleneck case and a rimmed or belted case when discussing 'Improved" chamberings.

Rimless and rebated cases are the case designs which <u>always</u> require a barrel set back to be properly headspaced when rechambering. As mentioned above, headspace on an Ackley Improved rimless or rebated designs is .004" shorter than standard. This shorter headspace means you have no choice but to set the barrel back if you want correct headspace.

Note the mark on the shoulder at right, where the factory case is touching the improved chamber.

27

The only place the factory cartridge case will touch in the new chamber that matters is the bolt face and the junction of the neck and shoulder on the case. It will actually slightly crush the case shoulder when you close the bolt on the factory round. If you eject such a case unfired you will normally see a shinny area on the shoulder where the case was crushed just a little. This crush fit maintains proper headspace during the fire-forming process.

Ackley commented on forming brass and possible problems if the headspace is not correct; "When fire forming new cases, separation troubles may not appear the first time a case is fired but there is a weakness created the first time a case is fired, unless the headspace is sufficiently tight to create a crush fit on the unformed new case."[2]   In a letter to another gunsmith Ackley wrote, "When a factory cartridge is chambered in the Improved chamber, it should require some force to close the bolt. When the empty case is extracted you can see a definite ring right at the base of the shoulder where it contacted the chamber."[3]

In recent years several of the commercial reamer makers have decided to offer Ackley gauges with the "Improved" shoulder angle (40 degrees is the most popular version). This change represents one more way that novices can be confused about the headspace on Ackley Improved cases. Such gauges are designed to utilize the SAAMI style datum line on the shoulder, in stead of the traditional Ackley method of the neck/shoulder junction. However, all such gauges this author has seen still provide the same headspace measurements, so the finished product is the same as Ackley intended. It's just a different way of approaching the measurements to get the same result.

Chambering an Improved Rimmed Case...

Rimmed cases are the easiest of all improved or wildcat cases to chamber for. The rim is the headspace control feature on these cases. The rim is trapped between the bolt face and the rim cut in

---

[2] Guns & Ammo, Q&A, P.O. Ackley, February 1967
[3] Letter to Bevan King, October 23, 1974

the back of the barrel. So if the rim is headspaced correctly you can have almost any shape of case fire-formed beyond the rim, so long as it will extract.

Rimmed Headspace Gauges pictured at left.

There is <u>no need</u> to set the barrel back on a rimmed cartridge when you convert it to an improved design, unless there is excessive headspace in the factory chamber. Why? Because the rim controls headspace, the fact that the shoulder will be moved forward and the neck shortened has exactly <u>*NO effect*</u> on headspace for rimmed cartridges.

Reamers for rimmed cases normally have the rim cutter integral to their design. Simply paint the rim cut in the barrel with machinist's blue (Dykem®), when the rim cutter gets close to this material just watch close, as soon as it scratches the material, stop reaming. Utilizing this method there is no danger of changing the headspace of the gun in the process of 'Improving' the chamber.

Chambering an Improved Belted Chamber...

What was said of rimmed cases above is also true of belted cases. Belted designs headspace on the belt much the same way rimmed cases headspace on the rim.

At Right is the Ackley Double Shoulder magnum case. This was an experimental design, but it illustrates that the belted or rimmed case can have extreme shapes and headspace normally.

The distance between the bolt face and the belt cut in the barrel is the headspace for these cartridges. Like the rimmed designs, 'Improved' belted cases use the standard headspace gauges, no custom gauges are needed.

While it is possible to use machinist blue as suggested with the rimmed case, you will quickly find that it is much harder to determine if the machinist blue has been scratched, there is simply much less area to view. For this reason it is a good idea to set the barrel back when doing a belted magnum improved case. If the barrel is slick with no sight holes drilled in it you can set it back so that that bolt will not close on the go gauge. Then rechamber with the Improved or larger cartridge design reamer until the go gauge will allow the bolt to close normally.

If you would like to hide the factory markings on a slick barrel, just set the barrel back a about a half turn so that the markings are under the water line of the stock.

Above: Belted Magnum Gauge

If your barrel has sight holes it will be need to be set back a full turn to align or "time" the barrel with the receiver properly, otherwise your sights will not be at top dead center. Once the barrel is set back you can simply rechamber to correct headspace.

280 Ackley Improved (Nosler)

This cartridge has produced an on-going controversy. When Nosler took the cartridge to SAAMI they used adapted the standard method of measuring headspace as with all SAAMI cartridges. Long story short… There is no difference between the reamers, and never has been, only the headspace gauges were in question.

The SAAMI or Nosler gauges deliver the same length chamber as the "Traditional" Ackley gauges. For liability reasons it is still wise to use the gauges requested by the client and mark the gun accordingly. I use the terms "280 Ackley Improved, Nosler" or "280 Ackley Improved, Traditional" when marking calibers on these barrels, just to keep the lawyers happy.

To see the ongoing discussion about the 280 AI/Nosler this web site is a good source:
http://gunsmithtalk.wordpress.com/2010/01/13/280-ackley-improved-alert/

The first book in the Gunsmithing Student Handbook Series, "Chambering for Ackley Cartridges" provides a considerably more in depth discussion of Ackley Cartridges and the specifics of working with and headspacing them, including the 280 AI.

To Gauge or not to Gauge?
Liability is the question.

I talk to hundreds of gunsmith's each month through my business. It surprises me how often these "professionals" decide not to use a headspace gauge when they are available to them. They say, "No, that's OK I will just use brass."

The reason that we are still using brass cases in firearms after more than a century is that so far brass provides the best mechanism for safely sealing the breach during the firing of the round. Brass is surprisingly forgiving and acts a bit like an inflatable gasket when pressures rise. The properties of brass have saved many an experimenter who did not understand the problems of headspace and pressure.

On some rare occasions you may not have a choice with regard to gauges. Obviously, if the gauge is not available and delivery time for one from a reamer maker or a rental agency is too long then you may be forced to move forward. Using brass or ammo as the gauge to set headspace is expedient, not ideal. Nobody likes a client to be angry about delays, so we push ahead with an expedient method of headspacing...

The primary reason for this discussion is that brass can and does vary widely in manufacturing tolerances. There are instances where the practice of using brass to set headspace is not really a danger or a liability. However, that does not mean that it's a good idea all the time. Starting with a list of the times it might be safe and reasonable to do so.

1. Rimmed cases
2. Low Pressure cases
3. Wildcat designs

With rimmed cases your pretty safe for two reasons. First, most are what we consider low pressure or are straight walled so pressure drops dramatically for ever increment the bullet moves down the barrel. Second, single shots and lever guns that normally accept these cartridges will function fine with a small amount of excess headspace.

A Simple headspace gauge takes all the guess work out of headspacing a chamber.

If we are talking about revolvers then using brass for headspace is something to be cautious about. It can certainly be done; however, I would check the rim thickness of my cases against the SAAMI specs for the cartridge at hand. If you have a minimum spec piece of brass and you set the gun to minimum headspace with it, then it will not be long before the client complains that the gun jams up and the cylinder will not turn.

If your client is a reloader or shoots commercial reloads the problem might be intermittent. Why is that? Because his brass is probably mixed from lots and makers so there seems to be no rhyme or reason as to when the problem appears. In reality the mixed brass is the reason for inconsistent results.

When we talk about low pressure cartridges we are referring to cartridges that operate under 40,000 PSI in rifles. Black Powder cartridges are good examples of this type of cartridge. The most popular of all is probably the 30-30 Winchester. Why is headspace less important with these cases? Because they will generally work safely with a large amount of excess headspace, simply because the brass is strong enough to deal with the low pressures generated. So your liability is minimal.

In the example of the 30-30 you can actually have so much

33

headspace in a 94 Winchester that the cartridge will sometimes misfire and yet you will never have a case head separation. That is because this cartridge when loaded to factory levels cannot overcome the strength of the brass to produce a catastrophic failure.

When you jump up in pressure though, as with the 375 Winchester compared to a 38-55 you will see that catastrophic failure is a real concern. Why? Because there is much more pressure in the 375. So using brass to headspace a midrange cartridge (40,000 PSI to 47,000 PSI) is a more delicate decision. Again you would want to be sure about the rim thickness of your sample brass vs. the SAAMI specifications for that cartridge.

With wildcat designs of your own making, headspace is totally up to you the maker. That is not to say it's not important. You should set a standard for any design you come up with. The reason for doing this is so that you can show that you work to specific dimensions if something were to go wrong. Plus it makes it much easier for you to diagnose problems if you know exactly what the dimensions are supposed to be.

If we are talking about an existing wildcat, then you should use the correct headspace gauge for that cartridge. If you fly by the seat of your pants you have no way of proving the gun was correct when it left your shop. Which brings up an important point, you should always test fire and retain at least one piece of brass for that gun. It makes your files a little fatter but it's worth it when you need to diagnose a problem when a gun comes back or you have to show that things have changed since you last saw the gun.

One way to handle the fired cases is to simply put them in a bag with the invoice number and file them on an annual basis with your other records. A box for each year with nothing but your exemplar brass does not take up much room and is easy to locate when you need it. It's habits like these that can save you time and money when working out a problem. Many times the client has changed the gun or damaged it in some way, or taken it to another gunsmith and these fired cases are a record of how the gun left your shop.

If there are times when using brass to gauge headspace is OK then there must be times when it's a bad idea.

1. High intensity cartridges
2. Wildcat cartridges
3. Ackley Improved cartridges

High intensity cartridges run from 47,000 PSI to 65,000 PSI. A high intensity cartridge with a little headspace can cause all kinds of problems. Excess headspace will allow brass to stretch at the web just ahead of the solid case head. This will lead to case head separations in as few as one loading depending on the dimensions of the chamber in question, and the dimensions of the brass in question. An astute reloader will set his dies to minimize case stretch, but you cannot rely on the reloader to understand this issue. I do recommend sharing information with your client to educate them on proper loading techniques.

If you have a client who loves to ride the wild edge of velocity and pressure (and we all do!) then keeping headspace to a minimum is an absolute necessity. When a cartridge is fired and excess headspace is present, or in other words, a gap between the case head and the bolt face, the action will take a pounding with every shot. On top of that, a hot rodding reloader is pounding the gun even harder. In some cases they are approaching a proof load in terms of pressure with every shot. You're not responsible for what they do, but you can still protect yourself from liability by keeping headspace as close to zero as possible.

Wildcats are often as mysterious to the gunsmith as they are to the shooter. There really should be no mystery. With wildcat cartridges using brass for headspace is a bad idea. As mentioned before brass can vary widely in quality and dimensions.

If the wildcat is of your own design, then make a gauge or have one made so that you're always working to known dimensions. If it's an existing wildcat then use the proper gauge. Over the years too many smiths have just done whatever they want for headspace. The problem is pervasive enough that the reloading die manufacturers now ask for a chamber print when they sell wildcat dies; even for old time well known cartridges.

"Quality Cartridge" is a company that makes custom brass for a wide range of wildcat and obsolete cartridges. Because of the failure of gunsmiths and hobbyists to hold proper headspace for Ackley cartridges, they ask for fired cases before they will supply formed cases for a client.

STANDARDS AND TOLERANCES...

The firearms industry utilizes voluntary standards. Small Arms and Ammunition Manufacturers Institute, Inc. (SAAMI) is the library and registry for these standards. Utilize the data SAAMI provides, it is invaluable. SAAMI does not register wildcats. Only members of SAAMI can submit cartridges for standardization. However, they will accept prints of your wildcats as a record and this could potentially save your bacon some day if you plan to market a cartridge actively.

The voluntary standards available from SAAMI protect us all from liability if we stay within them. Gauges are your best defense against liability issues, so whenever possible; use them.

You can download the standards from SAAMI at:
http://www.saami.org

When you read the specifications for a particular cartridge you will note that there are tolerances called out for cartridge and chambers. Take the time to look them over, become aware of the fact that there are acceptable standards already established. There is no reason for the gunsmith to fly by the seat of his pants. Even when working with a wildcat, many of the SAAMI standards for the parent case will still apply. Once you understand these tolerances you will be better able to diagnose problems with chambers and headspace much faster and more accurately.

I have seen articles that claim if a military gun is headspaced with SAAMI spec gauges that the results are suspect. Then the writers go on to point out variances of amounts in specifications of as little as .0015" or about 1/3 of the difference between a go and a no-go gauge in most calibers. Clearly such small differences will not get you in trouble.

Military or foreign specifications and tolerances will vary from SAAMI. For this reason it is wise to have the right information in front of you rather than making assumptions. An excellent example is the 7.62mm x 51/.308 Winchester cartridge:

308 Winchester Vs. 7.62 Nato

CARTRIDGE	MIN	MAX
SAAMI (308 Winchester)	1.627	1.634
GOVERNMENT (7.62 x 51 NATO)	1.630	1.633
CHAMBER	MIN	MAX
SAAMI	1.630	1.640
GOVERNMENT M-14	1.6355	1.6385
GOVERNMENT M-14 Match	1.631	1.633

Note that the government ranges are for new guns, while the SAAMI range allows for wear.[4]

In the above example it is clear that 7.62 Nato ammo would easily fit in a SAAMI chamber. Conversely, some SAAMI ammo would be rejected by the Government because SAAMI standards allow for a wider range both in minimum and maximum measurements.

So, if you have a firearm chambered for 7.62 you must use Government spec. gauges to check headspace, not 308 Winchester gauges. Naturally this presents the most problems for semi-autos like the M1-A and the AR-10 where SAAMI ammo might cause function problems. With all that said, bolt guns will likely digest either ammo without much concern.

Stupid Question Department:

"These gauges are wrong; they are marked go and no-go but I measured them and they are the same exact length. The tool maker must have made a mistake, right?"

I have actually had this question more often than you would expect from "professional gunsmiths". Adding the word professional to your job title does nothing to train you or increase your knowledge.

---

[4] Armalite, Inc., Technical Note 69, 2008

"Professional" just means you take money for whatever service you provide, much like a prostitute. Training and education are required to improve your understanding of any subject, knowledge far outweighs titles (like professional or master gunsmith).

No the tool maker did not make a mistake. You're measuring the overall length of the gauges, not the length to the datum line. Because shouldered headspace gauges utilize the datum line along the shoulder to set the length of the chamber a ring gauge of the correct diameter is required to measure the gauge.

 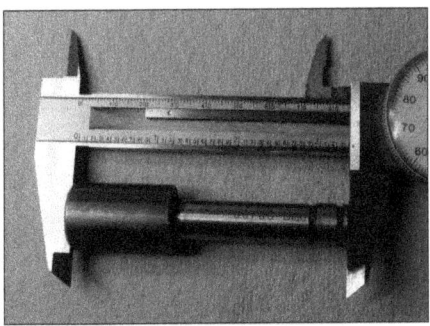

Simple ring gauge you can make. Just subtract the length of the ring gauge from the overall measurement and you have the length of the headspace gauge to the datum line.

Example: in the case of a 25-06 headspace gauge the datum line is at the point along the shoulder that measures .375". So, the length from that point along the shoulder to the back of the gauge is the "length" the gauge is built to measure. A piece of steel faced square with a .375" hole through it can be used to check the length of the gauge (as seen above).

Another area where I sometimes have gunsmiths show their lack of knowledge is when they order tools for two standard belted magnum cartridges and then order a set of gauges for each cartridge. One gunsmith went so far as to tell me that one cartridge was longer than the other so he needed different gauges. Sorry, standard belted cases all use belted magnum gauges. The couple of exceptions were covered earlier in the book when we discussed headspace methods.

Since confusion over gauge interchangeability is common. Included on the pages that follow is a chart the shows most popular cartridges and the chambers they interchange with.

Forster makes a tool that can be used to measure gauges or ammunition. This can be a very useful tool to diagnose headspace in customers reloaded ammunition.

The Ackley line of improved rimless bottleneck cartridges use the Ackley go gauge as a "go" gauge and the standard parent cartridge go gauge as a "No Go" gauge. Example: Chambering a rifle for the 30-06 Ackley Improved you would use the *"30-06 Ackley Improved Go Gauge"* as the go gauge and you would use the standard *"30-06 Go Gauge"* as the no-go.

SHIMMING GAUGES

The use of shim stock with a go gauge to create a no-go or a field gauge is not unheard of. It takes time and effort to shim correctly. If the shim stock gets wrinkled or doubled over along the edge of the bolt face the reading will be incorrect.

Shims for this purpose are normally cut from sheets of precision made shim stock. It is imperative that the shim be smaller than the bolt face so that it does not give a false reading. With all that said, shimming is a bad idea as a standard practice. It has too many opportunities for error. Whenever possible use actual headspace gauges made for the purpose to avoid errors.

The worst form of shimming involves the use of cellophane or masking tape. Tape is placed on the back of the gauge and trimmed to match the diameter of the rim. The tape can be measured to assure a reasonable accuracy. The problem with this process should be obvious; tape is pliable and relatively crushable. For this reason it is not a precise way to set headspace as a practice.

Even so, both methods of shimming can be used in a pinch to check the headspace condition of a chamber. Finesse and attention to detail are required for these methods to be used accurately. In the event that you choose one of these methods, carefully record the details of the shim thickness and the readings produced. These records may be essential if a problem develops later.

**Many headspace gauges are interchangeable.
Cartridges not listed in the pages that follow most likely have
their own unique gauges and are
not interchangeable with other calibers.**

Headspace Interchange Chart

GAUGE NAME	USE WITH THESE CARTRIDGES
Remington BR	20 BR, 22 BR, 6mm BR, 6.5mm BR, 7mm BR, 30 BR
PPC Gages	22 PPC, 6mm PPC
TCU	6mm TCU, 257 TCU, 6.5 TCU, 7mm TCU
17 Ackley Bee	25-20, 218 Bee
22 Long Rifle	22 Short, 22 Long, 22 Bentz, 22 Match, 17 HMII, 17 HS
22 RFM	22 Winchester Magnum, 17 HMR
22 Hornet	Hornet Based Wildcats
218 Bee	218 Mashburn Bee, 25-20, 32-20
221 Remington Fireball	221 Remington Fireball, 30-221, 221 Wildcats
223 Remington	6 X 45, 5.56 NATO, 17-223, 20 Practical, 223 Wylde
222 Remington Magnum	6 X 47 Remington
22 Nosler	6/22 Nosler 20/22 Nosler
223 Winchester Super Short Magnum	243 WSSM, 358 Indy
6mm XC	22 XC
6mm Remington	244 Remington
6.17 Spitfire	6.71 Phantom
6.53 Scramjet	6.17 Flash
25-20	32-20, 218 Bee, 218 Mashburn Bee

GAUGE NAME	USE WITH THESE CARTRIDGES
6.5 x 47 Lapua	6x47 Lapua, 20x47 Lapua, 22x47 Lapua, 30x47 Lapua
6.5 x 47 Lapua Ackley Improved	6mm Long Dasher
6.5 Grendel	6mm Grendel, 6mm BPC, 6.5mm BPC, 264 LBC-AR
6.5 Creedmoor	6mm Creedmoor
26 Nosler	28 Nosler (all other Nosler calibers have specific gauges)
7 X 57	257 Roberts, 6.5-257 Roberts, 270X57, 7mm Mauser
280 Remington	7mm Express
284 Winchester	22/284, 6/284, 25/284, 6.5/284, 270/284, 30/284, 338/284, 35/284
7.21 Firehawk	6.71 Blackbird
30-30 WCF	19 Zipper, 219 Zipper, 219 Zipper Imp., 219 Donaldson Wasp, 22/30-30, 25-35 WCF, 7-30 Waters, 7mm Int. Rimmed, 30-30 Based Wildcats, 30 Herrett, 303 Savage, 307 Win., 32-40 WCF, 32 Win. Spl., 356 Win., 357 Herrett, 375 Win., 38-55 WCF
30 Remington	32 Remington
300 Savage	300 Savage, 270/300 Savage
30-40 Krag	303 British, Krag wildcats, 40-70 Straight
308 Winchester	7.62x51 NATO, 243 Win., 260 Rem., 7/08, 338 Federal, 358 Win., 25 Souper
30-06	25/06, 6.5/06, 270 Win., 8mm/06, 338/06, 35 Whelen

GAGE NAME	USE WITH THESE CARTRIDGES
30 Remington	32 Rem.
300 Win. Mag.	See Belted Magnum
7.62 Patriot	8.59 Galaxy, 7.21 Tomahawk
300 RSAUM	270, 7mm and 338 RSAUM
300 Remington Ultra Mag	270 Ultra Mag, 7mm Ultra Mag, 375 Ultra Mag, 338 Edge (DO NOT USE with .338 Ultra Mag)
7.82 Warbird	8.59 Titan, 10.75 Meteor, 7.62 Firebird
32 H&R Magnum	32 S&W Long
8mm Mauser	8x57
338-06	30-06, 270, 35 Whelen
348 Winchester (WCF)	45-70, 450 Alaskan, 50 Alaskan
357 Magnum	22 Remington Jet, 256 Winchester Magnum, 357 Maximum, 38 Special
358 Winchester	308, 7mm/08, 243 Winchester
35 Whelen	30-06, 270, 338-06
375 Winchester	30-30, 38-55, 32 Win.
378 Weatherby	30/378 Wby., 338/378 Wby., .416 Wby., 460 Wby., 378 or 460 Based Wildcats
40-65	45-70, 348 Win.
44-40 WCF	38-40 WCF
44 Remington Magnum	44 Special, 445 Super Mag.

GAGE NAME	USE WITH THESE CARTRIDGES
45-70 Government	33 Win., 348 Win., 40-65 WCF, 45 Basic (45-120 3 1/4"), 40-82, 45-90, 45-110, 450 Alaskan, 50-70, 50-90, 50 Alaskan, 50-110, 50-140
45 Colt	454 Casull, 45 Long Colt
460 Weatherby	378 Weatherby, 30-378, 338-378, 416 Wby.
Std. Belted Magnum	.535" Base Belted Magnum Calibers: 257 Wby, 264 Win, 6.5 Rem. Mag, 270 Wby, 275 H&H, 7mm Rem. Mag, 7mm Wby, 7x61 S&H, 7mm STW, 300 H&H, 300 Win. Mag, 30-338, 300 Wby, 308 Norma Mag, 8mm Rem. Mag, 338 Win. Mag, 340 Wby, 350 Rem. Mag, 358 Norma Mag, 358 STA, 375 H&H, 416 Rem. Mag, 458 Win. Mag, 458 Lott.

If you do not see a cross-over gauge listed here then it is very likely that the cartridge utilizes a specific gauge that has no alternative. Gauge and reamer makers are a great source of headspace information when you are unsure.

About the Author...

Fred Zeglin, has been building custom rifles for nearly 30 years and specializes in wildcat designs for his clients. He has taught NRA Gunsmithing courses in Wildcat Cartridge Design at Murray State College in Oklahoma and Trinidad State Junior College in Colorado. Fred also worked with AGI to create a Wildcat Cartridge lesson on DVD called "Taming Wildcats".

Fred has written articles for Precision Shooting Magazine, Guns and Ammo, and many others. He has hosted an award winning podcast about gunsmithing at:
http://www.stitcher.com/podcast/gunsmithing-radio
Fred also writes a gunsmithing blog, that can be found at:
https://gunsmithtalk.wordpress.com

Fred has published three other books, "Hawk Cartridges Manual", "Wildcat Cartridges, Reloader's Handbook of Wildcat Cartridge Design." and most recently "P.O. Ackley, America's Gunsmith"

Fred is the first gunsmith to tackle the subject of wildcatting in a book since P.O. Ackley in the 1960s. The Wildcat book includes historical references as well as information about making reamers, manufacturing your own reloading dies, how-to, dimensioned drawings, and even some load data.

"Understanding Headspace" is part of a new series of gunsmith manuals that Fred has undertaken writing. Titles include Chambering for Ackley Cartridges, Relining Barrels, Glass Bedding Rifles for Stability and Accuracy, and Chambering Rifle Barrels for Accuracy. With more to follow.

Andy Hill at Hawk Bullets, had this to say about Fred, "During the normal course of business we have gotten to know some gunsmiths with superb skills, artists crafting metal and wood into fine and functional firearms. Usually their ballistic knowledge is well rounded, but we believe one such gunsmith is quickly becoming a modern day P.O. Ackley. He is Fred Zeglin and he has done extensive development of a line of wildcat cartridges gaining popularity for their ballistic properties and low felt recoil."
Fine praise from a craftsman of quality bullets.

This book is part of a series of gunsmith manuals that Fred is compiling. Titles include: Understanding Headspace, Chambering for Ackley Cartridges, Relining Barrels, Glass Bedding Rifles for Stability and Accuracy, and Chambering Rifle Barrels for Accuracy. With more to follow in the: "Gunsmithing Student Handbook Series".

Video Courses Fred has done for the American Gunsmithing Institute.

Hawk Reloading Manual

This hardback book contains 188 pages of stories, illustrations, anecdotes, instructions, and data.

Hawk Cartridges are unusual in wildcat circles in that, correctly headstamped brass is available for them. In partnership with Z-Hat Custom Inc., Quality Cartridge of Hollywood, NJ manufactures the brass as well as custom loaded ammunition.

Each cartridge covered in the book includes a dimensioned drawing. Contributions from Wayne van Zwoll, Michael Petrov, Dick Williams and Mike Brady are included. Pressure tested data is included for the majority of the load data and all loads are real world tested in firearms.

History of Hawk Cartridges is presented. This collection of data includes new material and new cartridges that were not included in the earlier electronic version of the manual. The intention is to provide information that time has shown to be valuable to shooters of Hawk Cartridges and for cartridge collectors.

You can buy the Hawk Manual @: http://www.4-dproducts.com
Or, you can be purchase it on Amazon.com
Hawk Reloading Manual

Wildcat Cartridges, Reloader's Handbook of Wildcat Cartridge Design

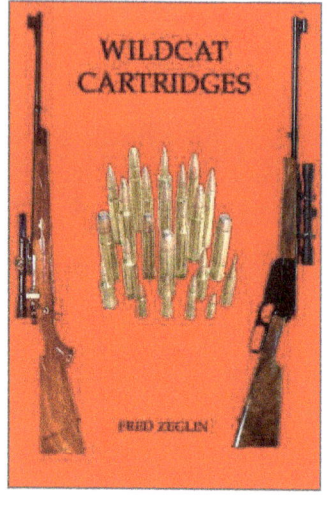

Wildcatting has been around almost as long as the metallic cartridge case. Wildcats have an air of mystery about them, no effort is made in these pages to diminish that mystique. Yet, you will find information here that is simply not available anywhere else. P.O. Ackley was the last Gunsmith to address the subject of wildcatting in depth. Over forty years later, Fred Zeglin, Master Rifle Builder and wildcatter has assembled in an easy to read, often humorous manual for anyone who loves guns, reloading, or wildcat cartridges.

History of wildcat cartridges is presented including many well known designers like P.O. Ackley, Jerry Gebby, and Charles Newton. The historical information provides an appropriate frame of reference for wildcatting. Nobody really wants to repeat something that has already been done. More recent wildcats are included along with reloading data and dimensions wherever possible.

Most valuable of all is the how-to information about making reamers and reloading dies. Fred supplies dimensions and instructions on how they are used to produce highly accurate reloading dies and chambers. Delivery times for such custom tools can delay a wildcat project by many months, knowing how to make your own dies can speed delivery of custom projects considerably.

Can be purchased on Amazon.com
[Wildcat Cartridges, Reloader's Handbook of Wildcat Cartridge Design](#)

P.O. Ackley, America's Gunsmith

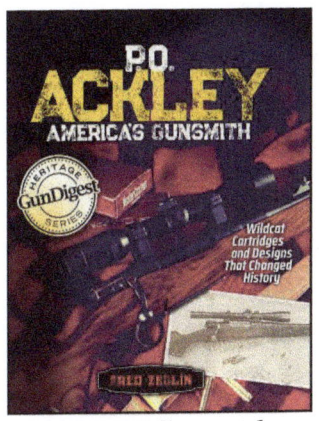

America's Gunsmith.

Parker Otto Ackley is arguably the most important gunsmith of the 20th century. He trained an incredible number of gunsmiths and shared a wealth of firearms knowledge along the way. The eminent gunsmith, ballistician, barrel maker, teacher and writer perhaps had more influence on modern shooting and firearms than any other single person. And now his life and works have been painstakingly detailed in *P.O. Ackley:*

Writer and gunsmith Fred Zeglin gives a never-before-seen look at the humble man whose research thrust the firearms industry forward. From pushing rifle chambers to their limits and developing superior barrels to designing red-hot cartridges, readers will walk away with a new appreciation for Ackley's exploration and ideas. And his concepts on reloading, rifle accuracy, safety, cartridge choice, and wildcat use are just as relevant for today's "gun cranks" as they were in Ackley's heyday.

Zeglin also delivers the most complete collection of accurate dimensions, loading data (much of it with pressure data) and history for the lifetime of cartridges created by P.O. Ackley.

Most shooters today know him because of his "Ackley Improved" cartridge designs. But those cartridges are only the tip of the iceberg. *P.O. Ackley: America's Gunsmith* is the whole story.

Bonus: Full-color photo section and an exclusive never-before-printed article by P.O. Ackley.

You can buy P.O. Ackley, America's Gunsmith @:
http://www.ackleyimproved.com

www.ingramcontent.com/pod-product-compliance
Lightning Source LLC
Chambersburg PA
CBHW040902250426
43672CB00034B/2986